BUGS UP CLOSE!

LADYBUGS UP CLOSE

THERESE M. SHEA

New York

Published in 2020 by The Rosen Publishing Group, Inc.
29 East 21st Street, New York, NY 10010

Copyright © 2020 by The Rosen Publishing Group, Inc.

All rights reserved. No part of this book may be reproduced in any form without permission in writing from the publisher, except by a reviewer.

First Edition

Editor: Elizabeth Krajnik
Book Design: Michael Flynn

Photo Credits: Cover, p. 1 Cornel Constantin/Shutterstock.com; (series background) Karuka/Shutterstock.com; p. 5 Achkin/Shutterstock.com; p. 7 (larvae) Geoffrey Budesa/Shutterstock.com; p. 7 (eggs) Jorge Abel Photography/Shutterstock.com; p. 9 Macrolife/Shutterstock.com; p. 11 InsectWorld/Shutterstock.com; p. 13 Henrik Larsson/Shutterstock.com; p. 15 Darkdiamond67/Shutterstock.com; p. 17 Martino77/Shutterstock.com; p. 19 Mario Saccomano/Shutterstock.com; p. 21 Ant Cooper/Shutterstock.com; p. 22 Valentina Proskurina/Shutterstock.com.

Cataloging-in-Publication Data

Names: Shea, Therese M.
Title: Ladybugs up close / Therese M. Shea.
Description: New York : PowerKids Press, 2020. | Series: Bugs up close! | Includes glossary and index.
Identifiers: ISBN 9781725307988 (pbk.) | ISBN 9781725308008 (library bound) | ISBN 9781725307995 (6 pack)
Subjects: LCSH: Ladybugs–Juvenile literature.
Classification: LCC QL596.C65 S43 2020 | DDC 595.76'9–dc23

Manufactured in the United States of America

CPSIA Compliance Information: Batch #CWPK20. For Further Information contact Rosen Publishing, New York, New York at 1-800-237-9932.

CONTENTS

All Around the World

Ladybugs are beetles found all around the world. Beetles are **insects** with hard wings that cover another set of wings. In North America, we often see red-and-black ladybugs. However, there are more than 4,000 species, or kinds, of ladybugs!

From Egg to Adult

A ladybug begins life in an egg. It comes out as a larva. Larvae eat other insects and insect eggs. Then, they become pupae, which are connected to leaves. Pupae grow into adults. Adult ladybugs **mate**, females lay eggs, and the **cycle** begins again.

eggs
larva

Parts of a Ladybug

Ladybugs are less than 1/2 inch (1.3 cm) long. They have a three-part body. The head has eyes, a mouth, and antennae. The thorax has three pairs of legs and two pairs of wings. The abdomen holds parts for **digesting**, breathing, and mating.

eye
antenna
leg
mouth

Wonderful Wings

A ladybug's outer set of wings is just a cover for the inner set. Its long, thin inner wings are called flight wings. Ladybugs aren't fast insects, though. They only fly about 15 miles (24 km) an hour.

Bright Beetles

Not all ladybugs are red. They can be yellow or orange, too. Their bright colors tell hungry animals to stay away. Ladybugs taste bad to animals! Some ladybugs have dark spots. Some have stripes. Some have no markings at all.

A Smelly Defense

A ladybug's colors aren't its only defense, or way of staying safe. If a ladybug is frightened, it can give off a bad-smelling yellow juice from its legs. Animals smell it and know to stay away! Their predators include birds, spiders, and frogs.

Ladybug Habitats

Ladybugs live in many kinds of **habitats**, including grasslands, forests, and towns. They're often seen in warmer weather. In cold weather, large groups of ladybugs look for warm places to sleep through winter. They might live under or in rotting logs, rocks, or people's houses!

Helpful Bugs

Most ladybugs eat other bugs, including those that eat plants. That makes them helpful bugs for farmers. Ladybugs eat pests, such as **aphids**, that eat farmers' crops. Some farmers even buy ladybugs and bring them to their farms!

Hurtful Bugs

A few ladybug species are hurtful to other ladybug species. Harlequin ladybugs came to North America from Asia in the 1980s. They may have a **parasite**, as well as **poison**. Both can kill native ladybug species that eat harlequin ladybugs' eggs and larvae.

Lucky Ladybugs?

Many people like ladybugs. In many places, seeing a ladybug is thought to be lucky. Some people just like ladybugs because most are bright and pretty. There are even ladybug toys! What do you think of ladybugs?

GLOSSARY

aphid: A sap-sucking, soft-bodied bug about the size of a pinhead.

cycle: A number of events that repeat in the same order.

digest: To break down food inside the body so that the body can use it.

habitat: The natural home for plants, animals, and other living things.

insect: A small animal that has six legs and a body formed of three parts and that may have wings.

mate: To come together to make babies.

parasite: A living thing that lives in, on, or with another living thing and often harms it.

poison: A matter that can hurt or kill people or animals if it gets in their body.

INDEX

WEBSITES

Due to the changing nature of Internet links, PowerKids Press has developed an online list of websites related to the subject of this book. This site is updated regularly. Please use this link to access the list: www.powerkidslinks.com/buc/ladybugs